PROPERTIES of MATERIALS

IS IT SMOOTH OR ROUGH?

BY LISA J. AMSTUTZ

PEBBLE
a capstone imprint

Pebble Emerge is published by Pebble, an imprint of Capstone.
1710 Roe Crest Drive
North Mankato, Minnesota 56003
www.capstonepub.com

Copyright © 2022 by Capstone. All rights reserved. No part of this publication may be reproduced in whole or in part, or stored in a retrieval system, or transmitted in any form or by any means, electronic, mechanical, photocopying, recording, or otherwise, without written permission of the publisher.

Library of Congress Cataloging-in-Publication Data is available on the Library of Congress website.
ISBN: 978-1-9771-3181-2 (hardcover)
ISBN: 978-1-9771-3282-6 (paperback)
ISBN: 978-1-9771-5451-4 (eBook PDF)

Summary: A piece of silk is smooth. Tree bark is rough. Why are some objects smooth and others rough? Find out in this series about materials and their properties.

Image Credits
Shutterstock: Africa Studio, 20, Alena Stalmashonak, 9, Apollofoto, 6, avtk, 11, BW Folsom, 14, Davydenko Yuliia, 4, de2marco, 19, (bottom), DJTaylor, 13, Gavran333, 12, Jiri Miklo, Cover, katueng, 5, LilKar, 19, (top), missanzi, 7, niwat chaiyawoot, 16, (right), OlgaGi, 18, RRandall, 15, Standret, 10, Tatiana Popova, 16, (left), turlakova, 17, vectorplus, (background) throughout

Editorial Credits
Editor: Michelle Parkin; Designer: Sarah Bennett; Media Researcher: Morgan Walters; Production Specialist: Laura Manthe

All internet sites appearing in back matter were available and accurate when this book was sent to press.

Table of Contents

Rough or Smooth?... 4
Friction 8
Changing Properties..................................... 12
Textures Everywhere 16

    Scavenger Hunt 20
    Glossary 22
    Read More.. 23
    Internet Sites... 23
    Index... 24

Words in **bold** are in the glossary.

Rough or Smooth?

Pick up a pineapple at the grocery store. How does it feel? The outside is bumpy. Bumpy objects feel rough. They feel uneven.

Next, pick up an apple. The skin does not have bumps or cracks. Your fingers can slide over it. It looks shiny. The apple is smooth.

Everything around you is made of **matter**. Matter has different **properties**. An object's properties tell us more about the object.

A **surface** can be smooth or rough. This is the surface's **texture**. Texture is how something feels.

Friction

Set a toy car on a smooth floor. Give it a push. The toy moves easily along the floor. Now try pushing the toy car on grass. It cannot go far.

Friction slows the toy down on the grass. Rough surfaces have more friction. Smooth surfaces have less.

The surface of an icy lake is smooth. It is hard to walk on ice. Your shoes slide easily. Boots with rough soles grip the ice. They add friction.

Cars can slide on icy roads. People put snow tires on their cars. The tires have extra bumps on them. Rough tires slide less on slippery roads.

Changing Properties

Rough objects can make other objects smooth. Sandpaper is rough. It has sharp edges. Wood is also rough. You can use sandpaper to **polish** the wood. The sandpaper's **grit** rubs away tiny pieces of wood. Now the wood is smooth.

sandpaper

Make your own sandpaper. Start with a piece of heavy paper. The paper feels smooth.

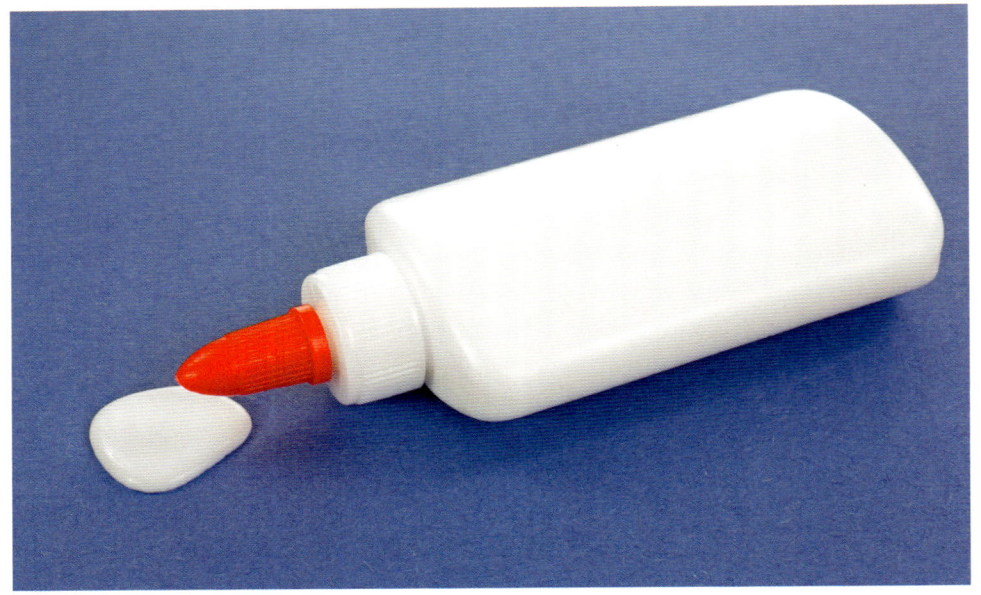

Spread glue all over the paper. Sprinkle sand over the glue. Then let it dry. Now your paper feels rough.

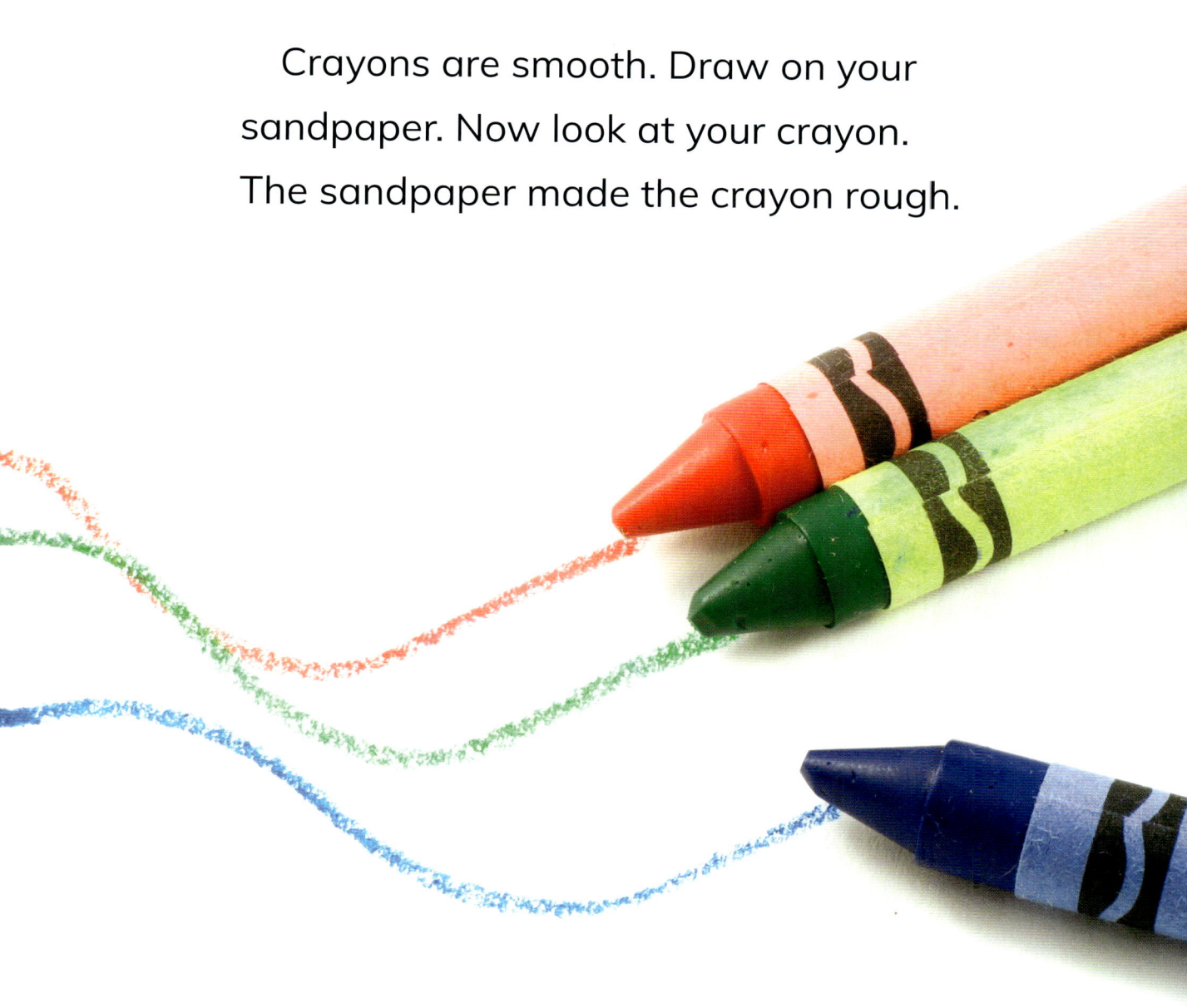

Crayons are smooth. Draw on your sandpaper. Now look at your crayon. The sandpaper made the crayon rough.

Textures Everywhere

A hockey puck is smooth. It slides quickly across the ice. A basketball is rough. It has tiny bumps on its surface. The bumps make it easier to grip with your hand.

A cat's fur is smooth. But its tongue is rough. The cat licks its fur to clean it. Its rough tongue smooths the fur.

What are you wearing today? Run your fingers over the fabric. Is it smooth or rough? Your jeans are bumpy. The texture is rough. Cotton and silk clothes are smooth.

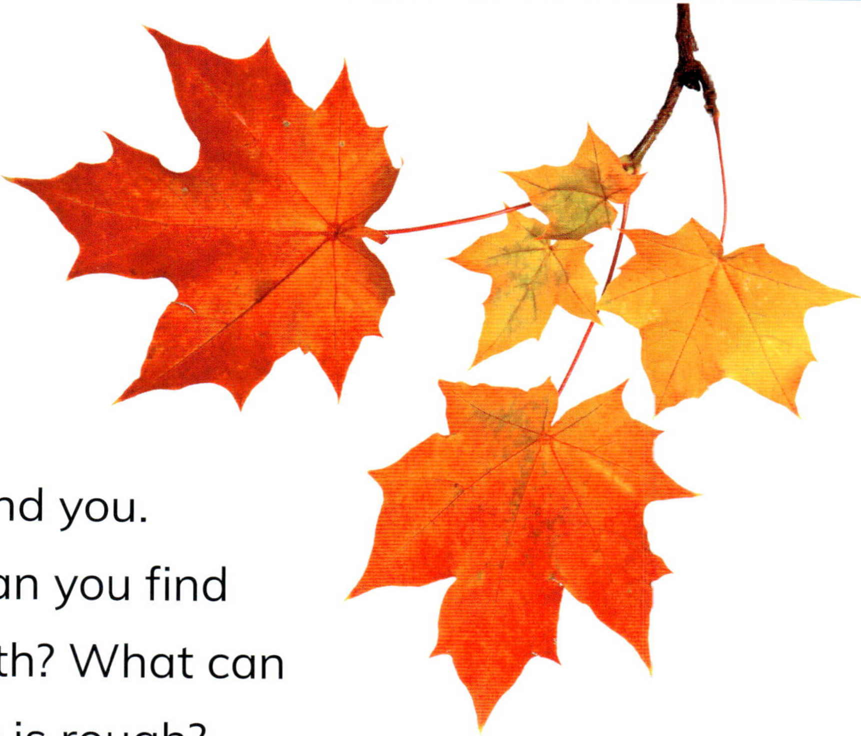

Look around you. What else can you find that is smooth? What can you find that is rough?

Scavenger Hunt

Find things around you that are rough and smooth.

What You Need:
- 10 small objects
- two baskets
- one piece of paper

What You Do:

1. Ask a parent or sibling to find 10 small objects around your house. They should have different textures.

2. Set out two baskets.

3. Write down the names of all the objects on the piece of paper. Go through your list. Is the object smooth or rough? Write down your guess.

4. Feel each object. Does it feel rough or smooth? Put all of the rough things in one basket. Put all of the smooth things in the other basket.

5. Look at your list. Circle the items that you guessed correctly.

Glossary

friction (FRIK-shun)—a force produced when two objects rub against each other; friction slows down objects

grit (grit)—very small pieces of strone or sand

matter (MAT-ter)—anything that has weight and takes up space

polish (PAHL-ish)—to rub something to make it shine

property (PROP-er-tee)—a quality of a material, such as color, hardness, or shape

surface (SER-fass)—the outside or outermost area of something

texture (TEKS-tcher)—the way something feels

Read More

Diehn, Andi. *Matter: Physical Science for Kids.* White River Junction, VT: Nomad Press, 2018.

Dunne, Abbie. *Matter. Physical Science.* North Mankato, MN: Capstone Press, 2017.

Kington, Emily. *Materials: What Is Stuff Made Of?* Minneapolis: Hungry Tomato, 2020.

Internet Sites

Friction
dkfindout.com/us/science/forces-and-motion/friction/

Physics for Kids: Friction
ducksters.com/science/friction.php

Index

apples, 5

basketballs, 16
boots, 10
bumpiness, 4–5, 16

cats, 17
changing properties, 12, 14–15
crayons, 15

fabrics, 18
   cotton, 18
   silk, 18
friction, 8, 10

glue, 14
grass, 8
grit, 12

hockey pucks, 16

ice, 10–11, 16

jeans, 18

matter, 6
   properties of, 6

pineapples, 4
polishing, 12

sandpaper, 12, 14–15
shininess, 5
shoes, 10

textures, 6, 16–17
tires, 11
toy cars, 8

wood, 12